The Cactus Name Game

by Barbara Wood

What do you see in this desert?

Cacti!

Many cacti are named after the things they look like. What do these cacti look like to you?

Let's play the cactus name game to learn some cactus names.

Look at each cactus.
Then guess its name.

This cactus has wide, flat pads. The pads are stems.

What's the name of this kind of cactus?

beavertail cactus

paddle cactus

lollipop cactus

It's called a beavertail cactus.

The pads are shaped like a beaver's tail. A new cactus life cycle can begin after a pad falls to the ground.

This cactus is round and prickly. It has many spines.

What's the name of this kind of cactus?

blowfish cactus

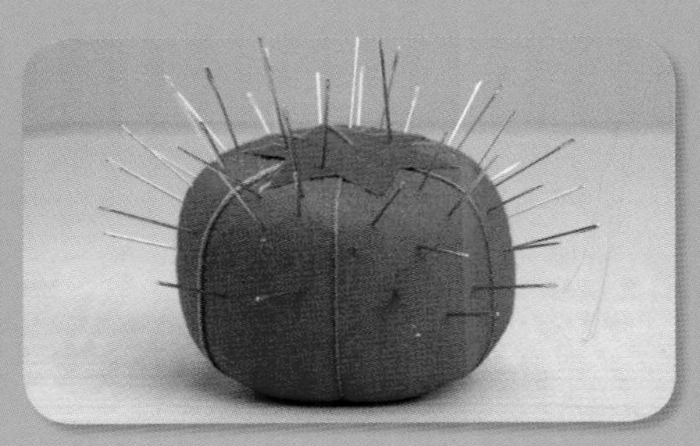

pincushion cactus

brush cactus

It's called a pincushion cactus.

The spines are like pins. They keep animals away. They also shade the cactus from the sun.

This cactus has a short, thick stem.

What's the name of this kind of cactus?

can cactus

lantern cactus

barrel cactus

It's called a barrel cactus.

The stem is shaped like a barrel. The stem stores water. The plant uses the water when there is no rain.

This cactus has many long stems.

What's the name of this kind of cactus?

spaghetti cactus

snake cactus

octopus cactus

It's called an octopus cactus.

The stems look like the arms of an octopus. Like other cacti, this cactus is named after the thing it looks like.